Canaries In The Sunshine

Other works by the same author

Fiction:
I Came Back
Ride a White Mare
Jade is My Stone
Splitting Apart
The Shades of Paracelsus
The Drain Brains

Nonfiction:
TAO: Total Person and One World
Tales of the Dragon, the Bear and Other Wondrous Creatures
Use Mind and Not Force: A Taiji Memoir

Verse:
Mostly Friday Nights
Pick up the Pearl

Canaries In The Sunshine

a celebration of

the coming century

of radiant energy

Patrick McGowan

Fomelhaut Publishing, Sydney, 2021

typeset by Patrick McGowan
using Libre Office opensource software
published by Fomelhaut Publishing 2021

ISBN 978-0-9925812-7-5

inquiries: pat@pjmcgowan.com
author blog: www.pjmcgowan.com
twitter: @_pjmcgowan

For those who believe
in the power of imagination

*Imagination is the translation of vibration
into something we can see.* Abraham-Hicks

Table of Contents

Not a Novel Idea........1
Want a Lift?........2
Us? Wrong?........3
The CAT That Got the Scream........4
If the Earth is a Battery........5
John Keely........6
The Hendershot Generator (HG)........7
Oliver Heaviside........8
They've Poisoned Me........9
200 Miles to the Gallon........10
Zip Plus Mileage........12
Air-Powered Car........13
Peter Perfect........14
Question of the Decade........15
Vandana Shiva Speaks........16
Must be a Record Somewhere........17
We Have Tremendous Spin Here........18
About Utron........19
The Crackpot........20
The Coler Convertor........21
The Lockridge Device........22
The Perpetual Motion Holder........24
The Gravitator........26
The Hutchison Effect........27
The Searl Effect Generator (SEG)........28
Skinner's Gravity Machine........29
The Eccentric........30
Viktor Schauberger........31
They Dubbed it Supernatural Electricity........32
Pharos Company's Achievement........33
The N-Machine........34
Meeting Adam Trombly........35
A Life of Service........36
Nothing Magic........37
The School Girl Energizer (SGE)........38
The MEG........39
Villanelle to Jim........40
Everywhere They Look........41
Seike's G-Strain Energy Absorber (GSEA)........42
A Swift Life........43
Tachyon........44
Stefan Marinov........45
You Can Do It Too........46

The Casimir Effect....47
The VSG....48
Ad Astra Per Aspera....49
Maintain the Torque....50
Cosmic Induction Man....51
The Turning Point....52
The ReGen-X Generator....53
Ari De Geus....54
Aaron's Path....55
The Lindemann Motor....56
The Sticky Tape Scientist....57
The Tokamak Sonnet....58
Jensen's UDT....59
The Sweet Device....60
The Magna-Pulsion Engine....61
A Cleaner Day....62
The Space Energy Generator....63
The Greener Power Machine (GPM)....64
The Lutec 1000....66
The Testatika Machine....67
The Banks Heat Machine....68
The Monothermal Power Device....69
The Adams Triplex....70
Something Good for Humanity....71
The Gyroscopic Particle Theorist....73
The Crystal Energy Man....74
The Seven Pillars of Tiller's Bridge....76
Hans Alfred Nieper....80
The Beam Ray Machine....81
Tesla's Advice to Walter Russell....82
The Multiple Wave Oscillator (MWO)....83
Universion....84
Science Cackle....85
Alan Holt....86
Orgone Matters....87
Cater Uncomplicates....89
One plus one....90
A Futurist Could See It....91
A Dark Matter....92
The Western White House....93
Boyd Bushman....*94*
Sestina on Pseudo-Science....95
Unpatented Words....97
Up from New York....99

Not a Novel Idea[1]

Ere many generations pass,
our machinery will be driven by a power
obtainable at any point in the universe.

This idea is not novel.

We find it in the delightful myth of Antheus,
who derives power from the earth.
We find it among the subtle speculations
of one of your most splendid mathematicians.

Throughout space there is energy.

Is this energy static or kinetic?
If static, our hopes are in vain;
if kinetic-and this we know it is, for certain-

then it is a mere question of time for when men
will succeed in attaching their machinery
to the very wheel-work of nature.

---

1 Found poem: Nikola Tesla to the American Institute of Electrical Engineers, in 1891;

Want a Lift?

Dear young readers, I hope these pieces
give you a lift, a new sense of power.
Though we touch on saucerology, please
realise the oomph's inside the metaphor.

Yes, we'll hear from the anti-gravity folk:
And some people think they're already woke!
But, hey, it's true we're caught up in words,
we wander the world in language herds.

Return mental blankness with timeless truth:
we're born creators, each carries the fire,
tweak and calibrate, the flame'll go higher.
But how about some words that soothe?

Haul in the anchor, set sail at your own pace.
The law of levity will always take first place.

Us? Wrong?

Lord Kelvin, or more correctly,
Sir William Thomson, First Baron Kelvin,
OM, GCVO, PC, PRS, FRSE,
first British scientist affixed
to the celestiality of the Lords,
coiner of terms like 'kinetic energy'
his cupboards and drawers crammed with awards,
known for his reach into every corner
of physics, mathematics and engineering.

From the peak of his achievements,
as if all the hard work had been done,
he scoffed at the new talk of aviation,
having not the smallest molecule of faith
in aerial navigation, other than balloons.

The CAT That Got the Scream

There's little worse than a case of clear air turbulence
to ruin one's day, ask those aboard Flight 911[2]
no visual clues, silent smiles all along the instrument panel
and the next minute, pilot and passengers are gone.

Over fifty years later, atmospheric science,
in free fall, still grasps at synoptic charts,
and theories of troughs, ridges, residual turbulence,
wind shear, jet streams, gusts and gravity fronts,

Dare introduce the concept of ether and related
gravitational energy field theory and the old science
wobbles, its stabilizers snap, it spins and nose dives
as it confronts its own disintegration.

As we transition to a new understanding of what air is,
hold onto your hats folks, we're in for a bumpy ride!

2 Refers to BOAC Flight 911 which crashed in Japan on 5 March 1966

If the Earth is a Battery

Farmer Nathan Stubblefield in 1888,
after years of testing, re-testing and sharing
with townsfolk in Murray, Kentucky,
patented his wireless telephone invention.

Wireless because he used the earth
as the conductive medium, the phones
were likewise charged with batteries
buried underground, a simple series

of coils and plates. The earth, he said,
is an electrical ocean upon which it receives
sets upon sets of waves, constant electromotive
force of commercial value, though they

fill with charge slowly – like new dams
(& much like the radiant energy movement itself).

John Keely

Author of *The Forty Laws of Sympathetic Vibratory Physics*, his life long passion-
the Keely Motor, fueled by an etheric force.
He confronted his skeptics with cool and calm.

One day twelve mining magnates challenged him
to demonstrate his disintegrator machine,
a research spinoff, up in the Catskill Mountains.

With a hand held tool, he disintegrated the quartz
leaving a tunnel eighteen feet deep, four feet wide.
Specks of gold sitting in the dust. To this day,
few can say, if it was about the yellow harvest for him,

skilled showman, or was he an anguished soul
struggling to reconcile with the simple truth,
we each invent our own reality – this we must decide.

The Hendershot Generator (HG)

Once upon a time Lester Jennings Hendershot
made world headlines with his fuel-less motor,
Further explained it wasn't a motor, but a generator,
magnatronic generator- the formal name given.

Lester spoke much of earth currents, triggered
by his early studies of the compass needle.
The device is technically a sensitive antenna
that picks up atmospheric electric frequencies

How can the earth be turning in a magnetic field
constantly, and yet no power be generated from it?
It's the cutting against the lines of magnetism
that leads to electricity, in the shed and on the earth.

Lester, in life, faced many moments of friction
which he struggled to reconcile as part of the plan.

Oliver Heaviside[3]

Although at odds with the scientific establishment for most of his life, Heaviside changed the face of telecommunications, mathematics, and science. Wikipedia

I remember my first look at the great treatise of Maxwell's
when I was a young man...
I saw that it was great, greater and greatest,
with prodigious possibilities in its power...

I was determined to master the book and set to work.
I was very ignorant.
I had no knowledge of mathematical analysis
(having learned only school algebra and trigonometry)

and thus my work was laid out for me.
It took me several years before I could understand.
Then I set Maxwell aside and followed my own course.
And I progressed much more quickly...

It will be understood that I preach the gospel
according to my interpretation of Maxwell.

3 Found Poem;

They've Poisoned Me

Stanley Meyer made a straightforward case
for running an automobile on water alone,
reckoned he could separate hydrogen and oxygen,
polarizing the elements through a resonance process.

Readers of today sadly spin on the one spot
as they seek to separate the truth from the lies:
some court in Ohio deemed his venture a fraud
so for many that was the end of the discussion.

Others say the case was tainted, and testify
that they had seen with their own eyes
his water-fueled dune buggy run, they also ask
difficult questions surrounding offers made
by vested interests, like one billion dollars,
for rights to his purported non-chemical technology.

200 Miles to the Gallon

I have nothing to gain by telling my story. If I knew I were to die tomorrow or if I were leaving the country, I would tell it...and it is quite a story. But I'm alive and this is my house. Charles Pogue

1

In the mood to clean his battered toolbox
Cornish mechanic, Patrick Davies, one day
levered the plywood floor, surprised to find
documents dating back to the thirties.

He told the reporter the toolbox was a gift,
a reward given to him forty years before,
after helping a woman paint her house,
said her husband spent some years in Canada.

Original plans for the Winnipeg carburetor,
vaporizing petrol before it enters the cylinder,
included an account of how the inventor
ran a lawn mower for seven days on a quart of petrol,

The perennial conspiracy chestnut reignited:
miracle carburetor that gives 200 miles to the gallon.

2

Woo hoo! Charles Nelson Pogue! 1933.
One day an inventor, up to the hilt in debt,
the next day, a well-paid air filter factory manager,
a tight-lipped lackey of the motor industry.

Work on a farm first fueled his interest in carburetion:
a gasoline tank that sat in a field, on a hot day,
exploded, to start him wondering whether vapor
rather than spray was how to feed a combustion motor.

He designed a unit, attracted attention, showing off
his modified 1929 Chrysler Imperial Roadster
giving him those two hundred miles to the gallon,
reckoned a few tweaks would get it to four hundred.

News of the event saw a plummet in oil stocks,
before the suits arrived to put it back in its box.

Zip Plus Mileage

A tale in passing of John Robert Fish (Bob)
and his home-made, multi-fuel Fish carburetor
designed on the principles of superior atomisation,
shingled the sales pitch: Zip plus Mileage!

Bob built his factory before the dirty tricks started,
every carburetor he mailed out was promptly returned
by US Post, stamped 'fraudulent product',
long enough to bust his business model.

Bob moved down to Daytona Beach, Florida,
supplied fishing boat owners, true early adopters.

It was in '55, Fireball Roberts drove a Buick
equipped with one of Bob's carburetors
in a pre-Daytona 500 race. Fireball came first,
by over a mile, but, surprise, surprise, got disqualified.

Air-Powered Car

A more probable scenario suggests that Neal traded the secret of the engine to the Germans to secure his daughter's release, and then abandoned any further work on it for the future safety of his family. Rexresearch

Details in his 1934 patent are assumed incomplete,
And his ageing son was vague on the machine's parts,
though he remembers how the Germans threatened him,
kidnapped his daughter, for full rights to the motor.

Bob Neal, of Arkadelphia, Arkansas, designed a motor
that ran on compressed air, researchers speculate
he was inspired by local legend or folk knowledge,
others say the idea was first presented by Tesla.

A simple operation: a series of valves that allow
low pressure air to enter the tank, with proper timing,
resistance to compression is absent, the air can be
hammered, the tank constantly remains at full pressure.

Imagine this: a self-filling air tank linked to a crank.
An air-powered car is back on the drawing boards.

Peter Perfect

Winning was nobody's but Peter Brock's,
as king of the mountain road, he endured.
One day released his energy polariser black box,
super tech that made 'a shit-house car' good.

Maybe he's a great driver, but this's a fraud,
this scandal's not a thing we'd ever applaud,
we can recognize a scam when it calls,
crowds condemned this sham box of crystals,

unaware, since the late seventies, it so arises
Jaguar's been installing a magnetic contrivance
on the fuel supply line to enhance performance,
ceramic tube magnets known as spin harmonizers.

Magnets are routinely applied to rocket exhausts
to enhance thrust! Yet one more of Brocky's firsts.

Question of the Decade

The question of *Who Killed the Electric Car?*
was more than a case of whether to plug in or fill up:
was a standoff between the cronies of big oil/big auto
and a growing citizenry with visions of a world free
of the petrol engine, the chance to clear smog,
make air clean again, reduce running costs,
build a motor with less moving parts,
cheaper cars with less calls for repair.

General Motors, churning ads for gas guzzlers
and roadsters boasting speed, style and power,
what ever the road condition, don't forget speed,
while elsewhere, fatherly-faced figures on TV
made their line on the EV perfectly clear:
let's not get too far ahead of ourselves, dear people.

Vandana Shiva Speaks[4]

So we are talking of the old oil economy
trying to maintain itself now with another raw material:
the green plant.

The only reason corn and soy has been planted
for biofuel in this country
is that subsidies make it profitable.

I think the big crisis of our time
is that our minds are being manipulated
to give power to illusions.

It's been shifted to measuring growth,
not in terms of how life is enriched,
but in terms of how life is destroyed.

---

4 Found Poem: Extract from Youtube interview;

Must be a Record Somewhere

In Oldenburg on Sunday, May 2, 1982,
an expert from the automobile testing institute
was invited to run a comparison test:

Motorbike A:
-standard commercial electric motorbike, with two batteries.
-reached a distance of 8.35 km.

Motorbike B:
-same type motorbike, equipped with a recharging tachyon energy converter;
-reached a distance of 36.3 km

Motorbike C:
-a standard commercial motorbike filled with 250 ml of gasoline
-reached a distance of 14.7 km

the report included the comment the excess energy of Motorbike B
comes from an 'aether field' at the present time not identifiable.

We Have Tremendous Spin Here

They doubted Otis T Carr's claims he could fly
his self-powered machine to the moon and return
all in a single day, but none could deny this showman's
word power, enough to carry a crowd to the moon.

Skeptic reporters persisted, wrote columns
on this 'shameless spouter of double-talk and baffle-gab'.
Crowned him king of the non-flying saucers,

His Utron energy accumulator– a self-charging battery
'uses the sun's magnificent force of electromagnetism
by means of natural reproductive chemistry.'

He went on and on, about man-constructed barriers,
about the dot at the centre of the crucifix
about imitating nature to become one with nature,
okay, okay, he flew some beautiful ideas in his time.

About Utron[5]

To me there's no such thing as a completed curve;
you only go half way, like you go half way into the woods,
then you're coming out. This is the same.

A bisection of a total sphere is its exact curve
and one half of it is primarily U-shaped.

Because in magnets there are always two poles
and one normal way to show them is in a U-shape,
but if it's in a bar magnet, there are still two poles
and the shape is still the same.

We can only put a rope one half way around the tree
and it's coming back the other way, and this is true in all wave motion.

Now, if you extend this into velocity,
this is the pattern in the sine wave
and also the electromagnetic wave.

5 Found poem: radio interview with Otis T Carr, rexresearch

The Crackpot

In a report on Moray, he was described by REA engineer Franklin P Wood as 'unfathomable' and an unmitigated ass, crackpot, nitwit, religious fanatic, essentially honest, and unable to originate or perpetuate a hoax or fraud.

Thomas Henry Moray invested twenty years
and two hundred grand on what he called
the Moray Radiant Energy Device
at its core, the solid state Moray Valve.

He called for serious, sane and systematic research
into this unlimited and free energy from the cosmos,
even wrote a book for the layman about his studies:
The Sea of Energy in which the Earth Floats.

Apart from the hundreds of first hand witnesses,
we can only presume his device worked,
If it didn't why would somebody have bothered to smash
it to pieces with a sledgehammer, as reported.

A sharp-eye surrounded by blindfolded cowboys,
another life passed by as excessive white noise.

The Coler Convertor

Neither Coler nor his backer were able to give any theory to account for the working of these devices using acceptable scientific notions. MI5 report.

German Naval Captain Hans Coler,
credible MI5 witness after the War:
'We lit our house with my self-powered generator
till the day the Allies bombed Berlin.'

I'd need a month to rebuild my Stromerzeuger machine
and a copy of my Magnetstromapparat,
(some said could power a submarine)
all I'll need is magnets, copper coils and condensers.

'The most important scientific principle
is that an electron should not only be seen
as a negatively charged particle,
but it's also a south magnetic pole.'

He built the machine, his interrogators, mystified,
wrote their report and filed it as classified.

The Lockridge Device

I
GI Lockridge throws new light on our subject,
His name is linked to a device that confounds many:
an over-unity energy machine, built from Volkswagen parts.

We meet him in Germany at the end of the war,
a US infantryman, a Screaming Eagle,
who scoured the country as part of the mop-up,
some say to support Operation Paperclip,
a drive to pin scientists and useful technology.

In a mountainous retreat, near the Eagle's Nest,
he and a GI mate, searching a house, find in the basement
a clunking motor, the source of light for every room.

No sign of power input, the motor runs without pause,
the only way to stop it is to jam the drive shaft.
They slyly pack the box in their personal belongings.

II
After a warm welcome home in Boise, Idaho,
Lockridge and his mate pull the box apart,
draw two sets of diagrams. And part ways.

Outsourcing the manufacture of machine parts,
to isolated workshops around town
Lockridge sells a three hundred watt free energy unit
to caravaners and campers passing by.

Business is good, gives more time for the beer,
which bothers his wife, who seeks love elsewhere,
a move our soldier finds impossible to bear.

Dousing his house and himself with petrol,
(a fuel his machine could have sent to the wall)
with a single click of his cigarette lighter,
his machine became buried in the darkness of time.

The Perpetual Motion Holder

I
Florida's famous fantasy rock garden
Coral Castle, built with one pair of hands,
those of eccentric builder, Edward
Leedskalnin, migrant from Livonia.

One thousand tons of finely carved stone,
each piece flush, not a crack of light,
all pushed into place by Ed alone,
its biggest rock weighs in at thirty tons,

there's an obelisk, megalith, water fountain,
astrology sculptures, chairs, table and throne,
a revolving gate weighing nine tons alone,
abstract works and a two storey home,

The work of one man, his hammer and chisel
plus wooden tripod, chains and a pulley.

II
The local legend is that he transformed stones
into weightless objects, though we're without
a single eye witness. Ed worked at night,
always alone. He welcomed daytime visitors,

who often asked him how in the name of hell
he did it, to which he would vaguely reply:
it's not that difficult if you know how,
and hinted he knew about the pyramids.
Addressing visitors, he often stared far away
and waxed about the name Sweet Sixteen,
who, they presumed, was a childhood sweetheart

he'd left behind, it could have been something else.

At the front gate, he sold self-published pamphlets,
including one on the nature of magnetic current.

III
This impecunious quarryman/stonemason,
in this pamphlet, claimed the foundations
of science as taught in schools are all wrong,
he dismissed the theory of electron and proton,

A double helix model of electricity better suits.
North and south poles are one and the same.
He talked of a universal everlasting substance
he named Cosmic Force, elsewhere Real Life.

The upshot was that he invented a device,
he named the Perpetual Motion Holder (PMH)
From here it's easy for a layman to get lost,
though he does outline a series of experiments.

In short, he viewed gravity as a magnetic force,
which his PMH could neutralize, hence the castle.

The Gravitator

Thomas Townsend Brown, of Zanesville Ohio
physicist, all through his career, migrated
from workplace to workplace, seeking to explore
his hunch that gravity was a form of radiation.

To the displeasure of many, he diverted funds
to feed his pet project on electro-gravitics,
and eventually assembled his Gravitor machine,
some call it gravitator, a disk-shaped object.

High voltage causes a weight differential
between top and bottom sides of such disks,
causing them to fall up. The disk divided
into segments could make it fall in any direction.

Today's new word is c*ounterbary*: the action
of levitation where gravity's force is overcome.

The Hutchison Effect[6]

A visit to John is definitely a once in a lifetime thing.

One evening in 1979,
Hutchison was in his laboratory,
sitting amidst the sparks
and high voltage effects of his equipment,
when he was struck on the shoulder by a piece of metal.
He threw it back and it hit him again.
This was the ignominious beginning of the Hutchison Effect.

In the ensuing months,
Hutchison found that by tweaking the settings on the equipment,
he could get things –ordinary household objects–
to levitate, move horizontally, bend, break and even explode.
In the latter case, this portion of the effect came to be known as `disruption'.

6 Found poem: Nick Cook, *The Hunt for Zero Point;*

The Searl Effect Generator (SEG)

Poor hearing from birth helped John Roy Robert Searl,
as he sought to share his idea with a world deaf to reason,
to find insight, into how best explain why he was here.
Spoke of recurrent dreams: hopscotch, steam rollers, ladders.
Forever asking: how is nature doing this?
He used numbers after he twigged to the law of the squares.
Discovered early why we should experiment outside.
Launched levity disks from the town common.

He declared he had learned how to churn the ether,
the man in the patent office described it as 'know-how',
told him to come back when he had a unit ready to go,
even as he was accused of stealing power from the grid.

He took to the road to spread his word: 'my magnetism
is not your magnetism, my magnetism is identical to light.'

Skinner's Gravity Machine

The simple fact is that we can generate electricity from a lever and an imbalanced wheel. We can apply this system in the same manner as any hydroelectric dynamo or facility, only we do not need water for a conduit. We never did. Nikola Tesla and the Niagara falls project was a patronizing jump back in time, as this could have been done far easier and far less expensively by using very simple mechanical systems that could have been engineered several thousand years ago. Kevin Hay

Perplexing archival footage: a gravity machine
built by William F Skinner of Miami.
A drunken contraption of metal rods and arms,
wobble weights, upper and lower, uncoupled

but connected, off-balance weights falling
all over the place to new centers of gravity,
consider it a mechanical amplifier,
massive torque from minimal input,

served by a motor of one eighth horsepower,
offers a twelve hundred per cent power jump
used to run a workshop, operate a lathe
to turn steel, cut half inch thick threads

William estimated this unit could sustain
the power needs of three thousand people.

The Eccentric

Another mechanical marvel,
the Cook Inertial Propulsion Engine,
invented by Robert L Cook of Vallejo,
turns centrifugal force into linear motion.

Off-centre rotators, on a common axis,
tapping a force some call the Coriolis,
the machine seemed to defy Newton's third law,
so testers refused to believe what they saw,

Cook noted it down in his *Death of Rocketry*
that trumpets the machine's enormous efficiencies,
a revolution in space travel is what he sees.
But an internal driving force reeks of eccentricity.

None know whether the powers bought Cook out
or he's a kook, undeserving our benefit of the doubt.

Viktor Schauberger

Implosion is no invention in the conventional sense, but rather the renaissance of ancient knowledge. Viktor Shauberger

Years in forests as a worker, wildmeister,
observations of pirouetting eagles and water flow,
over seasons, started a snowball of questions,
like, how does a trout climb this ten meter fall?

He went to the books and learned all the words
began writing about oligo-dynamic forces,
how nature is neither core nor covering,
how gravity is a densation of the force of levity.

Adamant that birds do not fly but are flown,
fish do not swim but are swum, it's water
which draws the fish forward, even that trout,
he talked of spirals, vacuums, and implosive energy.

In a Texan shed lie remains of an implosion device,
a vortex engine that runs on water, none can assemble.

They Dubbed it Supernatural Electricity

Welshman Ieuan Thomas, a man in the shadows,
oath-keeper and signer of the Secrecy Act.
Did it pain him to oversee the obsoletion of oil
and nuclear, as he loyally maintained his silence?

Notes from his company, Designex, reveal work
on a fuel-less self-powering negative electricity
generator system, modifications of an early design
of Shauberger's Repulsine Vacuum Motor.

The Merton Street building incident lives on in lore,
Toronto Hydro removed its fuse box and wiring,
attempting to shut off all power to the building,
though lights and appliances continued to work.

It's all there, handwritten notes and sketches made
in a face to face interview at the Lord Elgin Hotel.

Pharos Company's Achievement[7]

Since the end of the nineteen sixties
two North American laboratories have been working
in particular on gravity experiments.

The experiments led to levitation
(or a high increase of the gravity acceleration)
of material, including metallic objects
as well as dielectric objects such as styrofoam and wood.

Changes in the metric dimensions of metals do occur,
and the metals can be melted or shredded
in peculiar ways without being heated.

These experiments will play a role in defence technology.
They belong to the field of the still-to-be-mentioned Tesla weapons.

There exists a very impressive film documentation
for the aforementioned experiments.

7 Found Poem: from Dr Nieper's *Revolution in Technology Medicine and Society:*

The N-Machine

The seat of electricity is not within the conductor but is in the space surrounding it. Michael Faraday

Bruce Eldrige De Palma is our vanilla
example of a free energy inventor,
a physics teacher who attributed his design
to Faraday's homopolar generator disk.

We extract energy directly from space:
the primordial field of the universe only needs
to be distorted and energy flows in,
while a magnet can do this, we have many ways,

he says: even when we light a candle, the flame
derives its energy directly from this source,
the wax is consumed by the energy flowing through it,
all part of his molecular antenna theory.

In a state of effacement, he shared drawings and writings
that would inspire future generations to the Nth degree.

Meeting Adam Trombly

Meeting Adam Trombly has an impact that never seems to leave you.

It was a mighty first attempt, with Joseph Kahn,
fellow inventor, a closed path homopolar generator
that accessed electrical power from the air,
an output to input ratio of 4.92

The Invention Secrecy Act of 1951
was behind a visit to their lab outside LA
by thirteen paper-waving Federal marshals,
tasked with executing the gag order.

It was around this time that the lab was ransacked,
they dared not speculate who was behind the theft,
twenty three million dollars in replacement value.
some gear was later returned, though most of it smashed.

The only item they missed: a tube of mechanical drawings
of the generator design, stored in the false ceiling.

A Life of Service

A curious pattern within this research
into free energy inventors and their quest
for a machine that runs itself, I see emerge,
often our hero ends up second best.

Edwin V Gray Senior is a case in point,
'conversion element switching tube' he coined
for his eighty horse power gas-less automobile,
a car that recharges its own batteries.

'I had no schooling in electronics
so nobody told me it was impossible.'

This master of magnetic flux and the cold energy
source aroused ears on the heads of the DoD,
who convinced Gray he could better serve his country
through the manufacture of cold bazookas.

Nothing Magic[8]

There's nothing magic about free energy,
by free energy I mean something which produces output energy
without the need for using a fuel which you have to buy.

Free energy devices have been around for a long time now.
I have stood beside a water mill and its power is scary
as it could crush you in moments and never even notice.

That mill is on a quietly flowing small river
and it can operate at any time day or night
without anyone having to pay anything for the power it uses.

It will have cost quite a bit to build the mill in the first place,
but after that, it produces major power
day and night, day in day out, year in year out.

Most free energy devices are just like that
as it costs something to build them in the first place,
but after that they run for free.

8 Found poem: from Patrick Kelly website www.free-energy-info.co.uk;

The School Girl Energizer (SGE)

John Bedini was an inventor, not a scholar,
a man who knew how to throw words around
like 'this device is not an electric motor
it's a self-rotating energizer' or
'a photon is a fusion of energy and time.'

Yet he professed he could explain his work
to a ten year old school girl, one of whom
built her own free energy machine for first prize
at a school science fair competition,
now called the Bedini SG Energizer.

After a gutful of threats, he retreated
to his workshop to build miniature models,
research quietly, knowing one day his work
would light up the world, a spike in the Dirac Sea.

The MEG

Lieutenant Colonel Thomas E Bearden,
now 90, yet still hungry as ever to know,
far from his undernourished childhood days,
remains a man who lives outside the box.

In 2002 Tom and his team patented the MEG
(Motionless Electromagnetic Generator)
a key feature: nanocrystalline magnetic alloys.
Today the MEG remains motionless in the lab,

awaiting the nod from a risk venture capitalist.
the lab bench experimental model works well,
To commercialize requires some more fine work:
untangle the core's pulses to ensure they're additive.

Tom, inventor at heart, puts it like this: 'We've gotta
keep tweaking the hell out of it till it's just right.'

Villanelle to Jim

I can't speak now, I've got laryngitis,
a funny message for an answering machine.
Wrote I can't write, I've got writer's cramp.

When in a crowd, he gets all fidgety
Once a close friend of Bearden and Bedini,
He can't speak now, he's got laryngitis.

During the conference he seemed edgy,
After that was nowhere to be seen.
Wrote he can't write, he's got writer's cramp.

I'm the first to admit this story's sketchy.
Study these lines and read in between.
He can't speak now, he's got laryngitis.

This game has two sides and one is dodgy,
The other generates energy, cheap and clean.
Wrote he can't write, he's got writer's cramp.

Staying alive sometimes requires strategy,
I knew that woman, her name was Jeane.
I can't speak now, I've got laryngitis.
I can't write, I've got writer's cramp.

Everywhere They Look

You've got to be free in your head to be free in America. Eric Dollard

Oh, conspire, you conspirators!
And your conspiratees, conspiring too,
They conspire in co-conspiracy!
The conspiracist ideation is a conspiracy:
Even non-conspiracists are conspiracying too!

A conspiracy is a conspiracy is a conspiracy.
A sweeping conspiritism: I conspiracy,
you conspiracy, he conspiracies, she conspiracies,
we conspiracy, they conspiracy.

Conspiratorialism: a conspiracy of conspiritorialists
conspiritorially swallowed by conspiratology!
A conspiracy of anti-conspiratologists
conspire about conspiritorial conspirations
for a conspire-gate anti-conspirification.

Seike's G-Strain Energy Absorber (GSEA)

Professor Shinichi Seike penned it
in *The Principles of Ultra Relativity:*
how to build a transistorized coil device
that produces endless clean energy.

The big obstacle was how to convert
g strain energy, the Landau g factor,
into useable electrical power.

(Hats off to physicist, Lev Davidovich Landau,
hailed as father of Soviet tachyon research,
for his speculation on Landau oscillations)

Hector D Perez Torres, built a model
from off the shelf components, created
the rotary converter, aka RV (RotoVerter)
now open source, absorbing kicks from all angles.

A Swift Life

Gerald Feinberg, physicist turned futurist,
a man way out there, perhaps this is why
he penned the book *Life Beyond Earth,*
Kept asking, at seminars and dinner parties
what should our long range goals be?

Meanwhile he is accredited with the coining
of the word tachyon (tachys for swift)
to describe particles moving faster than light
Science refused to confirm their existence,
to which he shrugged his shoulders, and replied:

'It may be that tachyons simply do not exist . . .
[but] we may not understand why it should be so
until we reach a much deeper understanding
of the nature of elementary particles that now exist.'

Tachyon

They warded off the tachyon assault
on science by declaring it imaginary,
rendered hypothetical, they hoped to halt
revival of an archaic ether theory
that would topple the science/religion divide.

A core argument for the tachyon field:
Observations the speed of light is not constant.
A Kirlian snapshot of a leaf torn plant:
revealing past impressions of matter in space
added more substance to the tachyon case.

As the story of massless particles unfolds,
The bradyon, quark, boson, and so on,
we go beyond what's real, what's imaginary,
here in the field, we follow more what we feel.

Stefan Marinov[9]

After having walked so many years on the thorny way of truth,
I became tired. My books and papers are my scientific testament.
I hope that soon the absolute (Newtonian) space-time concepts,
which I restored by numerous experiments and by mathematical theory,

will be accepted by the scientific community
as those corresponding to physical reality.
I hope that the perpetual motion machines,
of which I constructed many prototypes

without closing the energetic circle,
will successfully be built by other people.
And if my achievements in space-time physics, in electrodynamics
and in the domain of the violation of the laws of conservation

will be silenced also after my death,
by leaving this world I can only repeat the eternal words: *feci quod potui*[10].

9 Found Poem; last statement by Stefan Marinov;
10 I have done what I could;

You Can Do It Too

So you see, you can build a fusion reactor with parts from an electronics store, auto parts store, welding shop, refrigeration supplier, hardware store, and craft store, perhaps with a bit of dumpster-diving on the side.... It really doesn't take tens of billions of dollars! Tom Ligon

Philo T Farnsworth, there to announce the arrival
of electronic television, circled by press and backers
for a public demonstration. 'When will we see dollars
from this invention?' One moneybag asked. In reply,
his very first image was a dollar sign. And so it goes.

In the mid-sixties, rid of the thermo mindset,
able to ensure inertial electrostatic confinement,
builds the world's simplest nuclear fusion reactor
in a lab funded by ITT who saw the huge potential
till a change of business heart. They cut funding,
refused him use of his patents they'd underwritten.
But the deed now done, today we read of teenagers
all around the world building fusion reactors
at their schools and in their homes. And so it goes.

The Casimir Effect

Colleges still fill students' minds with vacuum theory:
Absence of matter and energy, free space archetypal.
Some dare assert we see the vacuum unclearly,
inexplicable until the ether idea's back on the table.

Measurements show a force field exists if we
insert two uncharged conductive plates in a vacuum,
replete with particles and anti-particles, a rolling sea,
excitations on the screen- not what we assumed.

Finally we have proof- zero point energy is real,
fill your sail, new generation, with the ether wind.
Egyptian student, Aisha Mustafa, reckons she'll
connect her sail, with a patent on this new find.

The plan is to harness the dynamic Casimir effect:
Two shiny plates moving, with power to collect.

The VSG

Rene-Louis Vallee, engineer and nuclear physicist,
claiming to have proven the energy structure of space
and the electromagnetic nature of gravitation,
proceeds to build his VSG contraption:
The Vallee Synergetic Generator

There on the bench top in his lab
he constructs a nuclear fusion reactor
where a pure carbon rod is transmuted
to boron, releasing an energy pulse, before the boron
transmutes back to the carbon state.

With a few grams of carbon, that energy pulse,
could serve all of his country's electrical needs.

Think of it this way, carbon acts as a pump,
to express zero point electromagnetic energy.

Ad Astra Per Aspera[11]

Eugene Franklin Mallove, of Norwich, Connecticut
scientist, educator, publisher of *Infinite Energy*
magazine, drawn by nature to what's hard to explain,
like, that's what scientists do, don't they?

He was the first to agree, cold fusion research
still had plenty of wrinkles to be smoothed out,
but for the government to not only cut funding
but rescind grants absolutely stank. And he said so.

Refusing support for the most promising programs
only reinforced claims of the corruption of science.
From the cold fusion furore, came *Fire and Ice*,
and his next venture, the New Energy Foundation.

Gene gave it his best shot, before he instructed: 'Beam me
up Scotty, since there's no intelligent life down here.'

11 Through hardship to the stars;

Maintain the Torque

Paul Mario Babcock reckons there's
not a lot to it, electronics is much
like plumbing, though you're not working
with water, but the flow of electricity.

His Flyback Energy work becomes his story:
As he engineers high frequency switching
circuits able to eliminate back emf,
resulting in a reduced impedance effect,
he makes machines that draw radiant energy
with a high coefficient of power.

Commercializing products is a painful road
for most people, but he keeps his head down
and his spirits up, he's not one to prance
around and attract unwanted resistance.

Cosmic Induction Man

It's not easy to understand what he's doing
or where he's going, as he talks of integratons
and cosmic induction generators,

but Eric P Dollard is respected as one of the few
who has ever duplicated Tesla's experiments.
This electrical engineer scorns the free energy crowd,

says people have dabbled with magnets for too long,
is equally hard on mainstream electrical theory,
asserting they have only half of the picture,

bags thermodynamics as the cry of a death cult,
bent on diffusion, decay and dissipation,
while he reckons he can ignite new universes,

flames from which emanate new light, new music,
where science and life mutually serve each other.

The Turning Point

'I deciphered Tesla', Jim Murray begins.
Dogmatists fail to discern the difference
between energy and power, and energy can
be used many times over for an immense

output of power, with circuits made right,
manage the counter-electro-motive force,
prevent magnetism clashing with itself.
Helps to know we have two types of power.

The names of machines he invented are doozies.
undeciphered: toroidal rotating transformer,
the dynaflux alternator and the SERPS
(Switched Energy Resonance Power Supply)

This man who converts a motor's drag into thrust,
knows the drag science displays cannot last.

The ReGen-X Generator

Thane Heins, inventor, of Almonte, Ontario,
dropout, divorcee, a man dogged by detractors,
focussed for years on his 'Infinity Generator'
later christened 'Peripitea' (reversal of fortune)
got to thinking he had better reinvent himself.

He set up a company Potential Difference Inc,
updated his device, now known as ReGen-X
Generator, supported by bi-toroidal transformer,
(BiTT for short). With like-minded innovators,
offers a range of hydro-kinetic generator solutions.

All underpinned by the theory of replacing on-load
resistive effects with an assistive reaction instead.

Ari De Geus

Arie Melis De Geus, theorist, inventor
and author of *Fluidum Continuum Universalis*
suggests students should study ether physics,
because the future is in the non-material realm.

His research path was from hydrogen technologies,
fractional hydrogen that produces heat and energy,
to plasma fusion (with the help of a nuclear catalyst);
low energy transmutation forming plasma vortices.

He also sought to extract Zero Point Energy,
a primordial energy which can be calculated as
the product of the density of the fluid continuum (FC)
and the square of the speed of light.

After being so other-ized, he predicted his critics
would deny ZPE, even if they saw it with their own eyes.

Aaron's Path

Aaron Murakami, son of an officer
who travelled the world, trained to absquatulate
and resettle regularly, jet from base to base,
no longer buys into the culture of seniority,
especially within the scientific community.

Now pitched in one spot, this *qi gong* devotee,
researcher, author, educator and inventor
holder of patents on plasma jet injection systems,
and convener of national energy forums,
walks his own path, learns in his own way,

perhaps inspired by that *qi gong* maxim
of using four ounces to direct one thousand
pounds of force, which he seeks to achieve
both in the lab and in the flow of daily life.

The Lindemann Motor

The Lindemann rotary attraction motor
was but one of Peter's energy inventions
as he explored the world of pulsed engine designs,
and variable reluctance motors. To raise awareness,

he assumed the role of educator and author,
endeavored to compile a full list of reasons
why the Utopian state of Teslan free energy
hasn't arrived: Not only conmen of days gone by;

true, there's the money motive: while in theory
science smashes monopolies, in practice....oh well;
and national governments (in the name of security)
shudder at the thought another state may move ahead;

But most importantly, a non-demanding public,
waiting for a hero with their best interests at heart.

The Sticky Tape Scientist

This Howard Johnson, father of spintronics,
labored in a field that wasn't even a field,
developed a self-rotating, self-powering
magnet-based engine with no special name.

'Maybe I don't explain it well, but it works.'
Its rough circular arrangement of U-magnets,
one pundit commented, is oddly reminiscent
of that prehistoric site, Stonehenge.

Eyebrows became raised in scientific circles
at his treatment of the Bloch wall as a gate,
energy from the vacuum flowed without pause,
as he distorted north and south magnetic fields.

The soft spoken pioneer reckons there's enough
for the curious in *The Secret World of Magnets.*

The Tokamak Sonnet[12]

I have made a long enough descent into the Void to speak with certainty. There is nothing but Beauty- and Beauty has only one perfect expression, Poetry. And the rest is a lie. Stephane Mallarme

We shovel words into the tokamak machine,
home to an artificial sun whose
glow spirals as it grows across and down
from the title, a donut neatly spun,

illusions of gravity upon the page,
our emerging sun throws new threads
of light that twist from head to tail,
as the sonnetist checks each line for heat

whether it's the portable mini-model
or the Hefei HT7U, this gyroscopic
magnet-based contraption guarantees
last&firstword find quantum entanglement

warmth suspended within fourteen turns,
love dressed as a sonnet, and also a machine.

---

12 A tokamak is a device which uses a powerful magnetic field to confine plasma in the shape of a torus, it derives from the Russian acronym тороидальная камера с магнитными катушками (toroidal'naya kamera s magnitnymi katushkami) toroidal chamber with magnetic coils;

Jensen's UDT[13]

The time on Planet Earth is fifteen minutes before midnight; there is no time left to waste. Free-energy technology is not meant to be controlled by vain and greedy parasites who wish to use a gift from God to exploit their fellow man. Free-energy technology represents a spiritual transition of the human race. Free-energy is not meant to be owned, period! PR Jensen

I have built a transformer which supplies more power
to its load than is drawn from its primary source.
I named this device The Unidirectional Transformer (UDT),
because the magnetic reaction of the load current
does not affect the magnetic action of the primary circuit.

The UDT is composed of a parallel LC resonant primary,
a split secondary, a gapped magnetic core,
and a feedback winding, (writes little known Jensen)

I have invented and developed the UDT on my own,
without benefit of any knowledge of other free-energy devices,
if they exist, which utilize the basic principles of UDT operation.

Please feel free to use this information as you desire.
However, I hope that no one will attempt to patent
and control this type of transformer.

13 Found poem: Paul Raymond Jensen, letter to *Fulcrum* Dec 1994;

The Sweet Device

Floyd 'Sparkie' Sweet noticed things
over the years as a transformer specialist.
Sometimes fixed magnets would shudder
and vibrate, what we now call self-oscillation.

His designer barium-ferrite magnets
(thank you metallurgy) could be activated,
magnets that self-oscillate without end,
presto the Sweet Device, otherwise known

as the Vacuum Triode Amplifier (VTA),
essentially a gate through which energy
from outside can come pouring in.

This all works in accord with the theories
of energy from the vacuum, not to be found
in standard science textbooks studied today.

The Magna-Pulsion Engine

Benjamin Robert Teal of Madison, Florida,
whom we'll call Bob, wrote sci-fi for a hobby,
one story needed an engine, clean and silent,
so pleased with his design, he decided to build it.
He named this invention the Magna-Pulsion Engine.

Six electromagnets and a secret timing device
lays at its core, it uses no fuel and gives off no gases,
and is so quiet, sounds like a sewing machine.
Bob's sixth generation motor is large enough,
he says, to run a twenty ton conveyer.

The quantum leap in this electric motor design
is due to the way he recycles electrical energy.

Bob's moved on now, but I suppose he's pleased
with the story he shared about a cleaner, quieter world.

A Cleaner Day

Max Langenburg had free energy friends,
designed and built a few machines of his own.
This world he worked in, with not only its nutters,
was a cruel conjunction of science, big business
and men in dark glasses from the yonder complex.

Old science games were stagnant, a cesspit that stank.
Who wants to live waist deep in sewered slime,
bound by a thick muddy drag, with constant threats
to take you nowhere but where you stand, or down,
as you cling to rusted tools, and further feed the quagmire?

He started a company that cleans industrial liquid waste,
to produce high quality drinking water for everyday life,
the leftover sludge converted into useable energy,
to usher science back to its alchemical dream state.

The Space Energy Generator

The challenge to honor an epic life in a single sonnet.
Here's to Parahamasa Tewari, electrical engineer,
former director, Nuclear Power Corporation of India,
inventor, indomitable radiant energy researcher.

Mapped out his space vortex theory in books
before building his Reactionless AC Synchronous
Generator, more informally, Space Energy Generator
(the energy it 'generates' is plucked from space).

His bigger achievement was to convince his leaders
to include this energy source into the mix of the nation's
energy resources, a gigantic step, petrodollar countries
could never dare to imagine. Who knew Vedic studies

would one day throw such light on the relationship
between space, mass, inertia, light and gravity?

The Greener Power Machine (GPM)

I
He's often described as a good hearted soul,
Zimbabwean Sangulani Maxwell Chikumbutso
sayeth his inventions come directly from God,
hence the Saith Technologies company name.

This man from Africa, though he could be from Mars:
dropped out of school at fourteen to work on a farm,
later a long haul truck driver, never had a lesson
in maths or science, he was blessed with visions.

He claims radio frequencies are the energy source
for each of his Greener Power Machines (GPMs),
at their heart is a micro-sonic energy device,
that can run homes, cars, phones and drones.

Rigorous tests by those who study the laws
confirm the absence of other energy inputs.

II
He's made a lot of allies in science and business
circles. In passing, shares he's rarely met atheism,
most scientists he says believe in a God, indeed
science and spirituality must work together.

When outsiders ask how his machines work
he says he can intercept those radio waves
which are part of our loving earth's structure.
The finer details remain a trade secret.

He also proclaims his work will continue
even without him, because he has arranged

for his technology to be made open source
to help lift his African brothers and sisters.

Some find these two positions hard to reconcile,
though let's remember he's a good hearted soul.

The Lutec 1000

Two worlds set on different frequencies
would not normally meet, scientific fact.
but some feel compelled to make stuff happen,
often when bent on proving themselves right.

John Christie and Lou Brits from Cairns, Australia
two men with dirty big dreams, one day announced
their Lutec 1000 box would power a house
for free, after payment of the upfront fee.

The Australian Sceptic Society, avid proponents
of conformism to its own frequency,
scorner of diversity, laid out some bait
for John and Lou, who craved validation.

The boys quickly recalibrated to their own frequency,
left the resisters to resist, till the day they awake.

The Testatika Machine

Some say set an example- it's the best we can do,
for each has an antenna tuned to their own God.
Quite the view of the religious group, Methernitha,
living in Swiss Linden, and led by Paul Baumann.

A community powered by the Testatika machine
(consumes no fuel), an electrostatic generator
attached to an auxiliary electromagnetic circuit
of inductances, capacitances and rectification
that mobilizes the 'static' electricity it collects.

He's allowed scientists to observe the machine,
not out to convince anyone, he explains the design
is based on the study of lightning effects in nature:

The two counter-rotating discs generate a charge.
One disc represents the earth, the other the cloud.

The Banks Heat Machine

Ridgway M Banks of El Cerrito, California
learned early about the new memory metal,
Nitinol (nickel titanium alloy in equal measure).

Like an adept who never forgets his original posture,
a mini-robot sat in a rowboat, stroke after stroke,
enjoying real time spontaneous recovery, untiring,
after being bent and turned and pushed out of shape,
the model worker, focused on the task, ever resilient.

Ridgway moved through a series of prototypes on his way
to the Banks Heat Machine, not flash, not dangerous,
he unveiled the world's first solid state heat engine,
where waste heat is turned back into energy, and repeat.

Here's a science story that can't be distorted, though
rumors of a super Nitinol have been heard in some circles.

The Monothermal Power Device

Walter Lovell, inventor of the twist top,
his grand ambition: to make life easier,
presented the monothermal power device,
to convert ambient heat to electrical energy.

Two differing electronegative laminates
separated by a binding agent. Imagine
your smartphone constantly charged
once it's coated with these two layers.

The next step, a stand alone domestic unit
four cubic feet, stacked with layers,
able to power an average house,
without the need for any other input.

The man uncorked a world of possibility:
Why not monothermal panels rather than solar?

The Adams Triplex

Our universe is a sea of energy - free, clean energy. It is all out there waiting for us to set sail upon it. Robert Adams

Meanwhile, in the land of the long white cloud,
Doctor Robert George Adams, of Whakatane,
up on the North Island's Bay of Plenty,
lifelong student of new energy technology,
developed the Adams Triplex Aetheric Energy
Motor Generator, freely gave away his design,
its open source status ensured none could patent it.

The machine's success relies on its conversion
of particle spin (the undeniable nonstop motion
of subatomic particles- did someone say perpetual?)
into conventional electrical power. Adams accepted
the existence of an aether as a superfluidic
particulate medium which pervades all of space:
'The laws of physics...are only convenient assumptions.'

Something Good for Humanity[14]

I can do something good for humanity,
that's why I would never sell this invention
to the power brokers of the world,
and that means any son of a bitch that tells me:
'I don't trust you to do it, Joe Newman. I want to do it,
and you give it to me and let me make it, let me do it.'

No! Hell no! I've seen too many goddamned people who are goddamned liars
and it makes me want to crush their damned skull.

That's the way I feel inside,
and God has given it to me, he's kept me young so I can do it,
don't ever bring me anyone like that because it makes me dislike you,
are you smart enough to understand?

I'm sorry I ever met you,
because you lied to me,
you said that you'd never betray me,
when you do something opposite to what I want, you betray me!

I've had a lot of people who have betrayed me and betrayed humanity
and swore that they wouldn't.

I'll get this done without you,
I don't have to have you, I don't have to have anyone.
It ain't you who got me all this publicity back in the eighties,
and all the worldwide attention.

It was my work, my blood, and all you come along and suck ass
and say, like you're doing it all.

Well, you're not doing it, you're full of shit

---

14 Found Poem: Newman documentary by Jon Fox, 2015;

and you're stupid and you're crazy.
Let's just end it and I don't want you to call me again.

You understand that? I'm fed up with you,
you got fucken sense enough to understand
I don't want you around me again?

When you tell me shit like that
it makes me want to spit in your fucken face,
you ain't done shit, you dumb fuck,
you ain't nothing but a blob on the table,
you coward, goddamned son of a bitch,
don't tell me that you're a man better than me...

...and I can't get something done,
you ain't fit to wipe my damned shoes,

you think you can say something that'll piss me off and I'll take it,
but I won't take it, you idiot,

you must be fucken stupid,
now, get the hell out of here!

The Gyroscopic Particle Theorist

During Hurricane Betsy, from a mountain
top, he stared at the stars and pledged to God
from this day on, he'd be a tiger for humanity.
Set up a workshop in Lucedale Mississippi.

Some home study of magnetism and electricity
led Joseph Wesley Newman to believe
that within any magnetic field
there exists an endless river of energy.

Fiddling in his workshop, he added a paddlewheel
and presented to the world, in a series of steps
what papers called 'Newman's Impossible Motor'
output exceeded input by eight hundred per cent.

After bureaucracy, bankruptcy, divorce and breakdown,
this tiger dismantled his machine, rejoined his God.

The Crystal Energy Man

I
Pyramid Power, seventies bestseller,
trojan stroke of marketing genius,
by Gillis Patrick Flanagan, inventor,
and self-titled encyclopedic synthesist.

The book starts slowly, a take on energy
across a raft of cultures and civilizations,
includes a section on the pyramid of Giza,
before a magisterial account of modern physics:

how scientists today must re-admit
ether theory, that frictionless fluid,
a canvas of vortex rings, to explain
the true nature of energy and electricity.

From one who, at age seven, told his mother
he was the reincarnation of Nikola Tesla.

II
Life had been hectic from his teens, inventing,
writing patents, advising government agencies
like the Department of Unconventional Weapons
and Warfare, and driving Pentagon think tanks.

Before 30, branching out, wrote the pyramid book,
In one interview, Pat was quizzed about the danger
of educating black operators, he replied: 'that's why
I only share with them one per cent of what I know.'

Later invested fourteen years, as a techno-monk
studying water and other peaceful remedies,

formulating health and wellbeing products
that could better help people help themselves.

The ether issue still floats out there, a gathering cloud,
as Pat smiles down, waits for the water cycle to kick in.

The Seven Pillars of Tiller's Bridge[15]

My long-term goal in the psycho-energetics area since 1970 has been to help build a reliable bridge of understanding for humans that seamlessly joins the foundations of traditional science on one end, extends through the domains of the psyche, emotion, and mind and is firmly planted in the bedrock of Spirit at the other end. And further, that this bridge is both strong enough and reliable enough that, eventually, normal folks and traditional scientists will joyfully walk across it. William A Tiller

Pillar One

Eyes planted on an invisible sun,
Copernicus didn't evict the earth
from the centre, he saw a grander design,
heralded a plan for endless revolution.

Descartes, like a dog trampling reeds before rest,
practiced daily meditations and penned them,
driven by doubt, till from the luminous core,
he found the ground for a miraculous science,

Medicine has never been about the chemistry
it's the same old story of our angels
overcoming those demons in dark places
calling us away from what we know.

Today, people pay well for weekend workshops,
learn about the mind and the art of bending spoons.

Pillar Two
adaequatio rei et intellectus (the intellect must be adequate to the thing)

Dare put mind and brain into the same sentence,
and be warned: you're on a slippery slope,

15 A response to *Bridging Science and Spirit (The Genius of William A. Tiller's Physics and the Promise of Information Medicine)* by Nisha Manek;

Hours spent in laboratories,
breaking down the bottle, in search of the geni,

Neuroscientists, in the likeness of their own god,
wish to carve in stone the ten commandments of the brain,

but the brain is like an open field after fresh snow
(neuroplasticity) and how about their mind games,

unaware they're staring into a mirror
and intention is an essential ingredient,

and still cannot explain the joy we feel
sinking hands into fresh wet clay

yes, desire, the messy sandpit of non-science
where we discover brain is a verb and not a noun.

Pillar Three

The exam question asked for one thousand words:
Explain the second law of thermodynamics.
He wrote 'water flows downhill' full stop,
before the entropy of dissipated words
and ideas dragged him into a low energy state,
the lecturer decorated him with praise and high marks.
Nobody legislated those first and second laws
and nobody enforces them,
yet they provide endless work for physicists
who scratch and scribble like accountants and lawyers
to advise on loopholes and likely outcomes,
make projections for the arrow of time.

How do immutable laws explain the spontaneous,
or the enduring value of life's non-equilibrium?

Pillar Four

It started as an experiment about mind power,
exerting human influence on inert objects
from long distances, going double blind,
we set up an intention host device (IHD)
and an unimprinted electronic device (UED)
that's for control purposes, as science demands,
we also ensured no cross-talk between devices.

We dreamed big, but the modest scenario began:
can we influence water? We were curious
to know if we could affect its pH from a distance,
and yowza! The results came out consistently good,
totally repeatable events, that goddamned water
listened to our thoughts so well and was so intelligent
our university conferred it with an honorary pHD.

Pillar Five

What do we know about space?
It's not the empty backdrop we once imagined.

The case is building that space has structure:
it bends, in the presence of large objects;
it ripples, influenced by gravitational waves;
it even expands, much like a balloon;
And now we hear it responds to intention.

Space and counter-space is the newest take:
We have normal everyday D-space (direct space)
now add R-space (reciprocal space)
where properties display an inverse relationship
and are amenable to space conditioning.

Compare the numbness and aridity of the test lab
to the warmth and passion of the inventor's garage.

Pillar Six

Mark lives in Minneapolis Minnesota
suffers from severe arthritis, his doctors declared
they had run out of options, even for relief, so
he opted in on an information medicine experiment,
never met his doctor face to face,
they devised a statement of intention,
covering physical, mental and emotional goals.

The doctor, a believer in R-space, who conditions
his clinic as a place conducive to health and healing,
created a personalized intentional host device,
regularly re-activated, to make it meta-stable,
They described it as a cascading case of inside out healing,
at first, Mark had good days, though pain lagged in the process,
till one day he said:
'......for no apparent reason, the pain disappeared.'

Pillar Seven

What should we say to the lowly nucleus
at the centre of the hydrogen atom when we consider
if that atom was made the size of a sports stadium
this nucleus would be the size of a marble?

When I visit a river, mountain or forested valley,
that beams with beauty up to the sky,
and I tune to the subtle energies at play,
I can easily imagine the place being sculpted
by thousands of invisible hands,
the hands of those who once lived on this earth
and continue to shape the landscape,
their only tools being love and affection.

Dear nucleus, I want to be like you, eternally
satisfied, as I light up my own corner of the world.

Hans Alfred Nieper

Hans Alfred Nieper, internist, pioneer
of medicine (focus on health and vitality),
member, New York Academy of Sciences, co-founder,
American Association of Gravity Field Energy,
convener of a seminal symposium on energy technology,
author of several books on the subject,
exploring the link between the tachyon field
and the cells of a living organism.

His wikipedia entry notes his medical training,
and a warning his practices are discredited, unsafe.
No other interests mentioned, a link to the article
on quackery is kindly provided under the stub.

Ah, for the day this site is no longer guarded
by gatekeepers learnedly numb to the spirit of science.

The Beam Ray Machine

It has been reported that in 1980, the American Medical Association was found guilty by a US Court of Appeals of "conspiracy to restrain competition. . . New methods of health care have been discouraged, restricted and in some instances eliminated." Royal-Rife.com

For the first time, Rife's Universal microscope
allowed him to see microbes in their live state,
Each microbe having its own resonant frequency
he bombarded them with light of a matching range.

That banquet to honor Royal Raymond Rife,
and celebrate 'the end to all disease' was far too hasty
for the Beam Ray Machine would never
be a viable commercial proposition:

it was un-intrusive; drew no blood; caused no pain;
aroused little fear and left no scars;
few patients needed to return for follow up;
medicine was zapped harder than the pathogens.

Disclaimer: I make no claims this machine can cure,
but, man, it should've been studied and tested some more.

Tesla's Advice to Walter Russell

With Lao beside him, the artist explained:
light doesn't travel anywhere, rather,
it replicates itself. This is a universe
of One Thing. And matter separated
from this oneness, naturally seeks it.

He went on and on: it's all about balance;
balanced interchange; to be more accurate
let's call it, rhythmic balanced interchange.
Gravitation is the creative principle that folds
back to the seed, while radiation unfolds.

Nikola listened, his eyes, cannon ball of light,
'All true, but I suggest you bury your ideas
deep in a vault, and wait one thousand years
before human beings will ever understand you.'

The Multiple Wave Oscillator (MWO)

Lifelong advocate of the science of happiness,
New Yorker, Georges Lakhovsky,
studied the bio-electrical nature of life,
the electromagnetic capacity of living beings,

Led to believe health of the human body
is as simple as a well-trained orchestra
where each organ plays its signature tune,
guided by a lively though silent conductor.

Gave us the multiple wave oscillator (MWO),
a transmitter of radio-cellulo waves,
a flurry of waves that restore oscillations
to damaged cells within the organism,

a tool to restore cellular orientation,
retrieve the music, and presto happiness.

Universion[16]

I believe this ether is not the negation of all matter
but rather the synthesis of all radiating forces,
and therefore I have given the name Universion
to the universal plexus of all cosmic rays.

Universion is a conception of the infinitely great,
symbolized by the boundless universe,
and the infinitely small, the granule of electrified matter,
symbolized by the ion which is a world in itself.

The infinitely great of the universe is, in fact,
nothing but the integration of infinitely small ions.

Universion is ubiquitous and all pervading.
Every moment we have evidence
of its presence, as effective as it is silent.
The material universe and life itself are but unstable.

16 Found Poem: Georges Lakhovsky *The Secret of Life* page 92;

Science Cackle

I'm simply pointing out what has always existed. Corentin Louis Kervran

From an innocent farmyard observation,
chickens, scratching and pecking calcium-poor soil,
continue to produce hard-shelled eggs
came the claim of biological transmutations.

A theory that set the henhouse aflutter,
Cold fusion? Elements changing from one to the other?
He later rephrased it: low energy transmutation.
Spread the idea that plants feed minerals to the soil.

A cubic mile of ocean contains as much colloidal gold
as is stored in bank vaults across the world.
Why no sign of lead out there in the deep blue sea?
We now gnow ocean bacteria turn those elements into gold.

Beware! Ayatollahs of the immutable elements!
You think you won't be converted? Buk, buk, buk, bagawk!

Alan Holt

He wore his NASA hat, in '81, when speaking
at the first International Symposium
on Nonconventional Energy Technology,
subject: conversion of gravitational energy fields.

Promoted into NASA's entrepreneurial branch,
(commercially reimbursable payload projects)
he negotiated and advised on experiments
to be conducted on Space Stations and Space Labs.

Free energy believers certainly go places,
last I heard, he was a guest speaker at the 2020
International UFO Congress, his theme:
Visitors and Humanity's Future as an Interstellar Species

We once watched the sky for souls from past the perimeter,
Now we prepare to become that same visitor.

Orgone Matters[17]

He said it could be used against disease,
and as a weapon in the fight against the fascist pestilence.

I
Orgone energy is also demonstrated visually,
thermically and electroscopically,
in the soil, in the atmosphere,
and in plant and animal organisms.

The flickering of the sky,
which some ascribe to terrestrial magnetism,
and the glimmering of stars on clear dry nights,
are direct expressions of the movement
of the atmospheric orgone.
The hitherto misunderstood formations of clouds
and thunderstorms are dependent on changes
in the concentration of the atmospheric orgone...

This can be demonstrated by measuring the speed
of the electroscopic discharges.

II
The living organism contains orgone energy
in each one of its cells (says Wilhelm Reich),
and continuously charges itself orgonotically
from the atmosphere by means of respiration.

The 'red' blood corpuscles are microscopic,
orgone-charged vesicles having a blue glimmer;
they carry biological energy from the surface
of the alveoli of the lungs to the body tissues.

---

17 Found Poem: from Wilhelm Reich, *Function of the Orgasm;*

The chlorophyl of plants, which is related
to the iron-containing protein of animal blood,
contains orgone and absorbs orgone directly
from the atmosphere and from solar radiation...

The human organism is surrounded by an orgonotic energy field
whose range depends on the organism's vegetative liveliness.

Cater Uncomplicates[18]

There are many more complicated solutions to a problem.
In actuality, the number of solutions is generally
in direct proportion to their complexity.

This means it is more difficult to find a simple solution
because there are fewer of them.

The degree of intelligence required to find a solution
to a given problem is inversely proportional
to the amount of additional information needed,

over and above the minimum number of facts
which are necessary in order to solve the problem.

Therefore, the simplest solutions are generally
the most difficult to find unless one is lucky.

In any event, the simple solution has the greatest likelihood
of being correct since nature's laws are basically simple.

---

18 Found poem: Joseph H Cater *Awesome Life Force;*

One plus one

He used to dare think that as a general rule whenever we have one and then one more, we also have three, and five, and seven, and infinity.
Velimir Khlebnikov *Let Them Read on My Gravestone*

What's more elementary than arithmetic?
It's an operation anyone can do,
a stick in this hand and in that hand, a stick,
one plus one always equals two.

But as I gaze out my window at the apple tree
I conjecture how one plus one can also equal three,
For each plump apple houses so much seed,
Our monopolistic number theory must concede.

If I ask one or more persons to join with me
to form a team for high-mindedness or fun,
We soon discover one plus one still only equals one,
each of us an island upon that single sea.

How can we count as authentic human beings,
as we navigate the gulf between thoughts and things?

A Futurist Could See It[19]

The Mussorgsky of the future
is giving a coast-to-coast concert of his work,
using the Radio apparatus to create a vast concert hall
stretching from Vladivostok to the Baltic
beneath the blue dome of the heavens.

On this evening he bewitches the people,
sharing with them the communion of his soul,
and on the following day
he is only an ordinary mortal again.

The artist has cast a spell over his land;
he has given his country the singing of the sea
and the whistling of the wind.

The poorest house in the smallest town
is filled with divine whistlings
and all the sweet delights of sound.

---

19 Found Poem: Velimir Khlebnikov, *The Radio of the Future;*

A Dark Matter

We're blinkered by words, often self-imposed,
when we define a problem in unhelpful ways,
even our scientists, hard-headed, hard-nosed,
consider 'dark matter' a convenient catchphrase.

What, pray ask, are we expecting to find,
as we reach in the dark, voluntarily blind?
So deep in the forest, we can't see a tree,
and imagine science will avert the catastrophe.

Their river of muddled words flows lava-like,
we break up dark matter into hot, warm or cold.
There's also dark energy, as though we can't hold
a torch to it, no instrument able to ping or strike.

The great pity is, a quick brush up on metaphysics
could save us millions resolving this word fix.

The Western White House

George Walker Bush, immortalized
as forty-three, only ever dabbled in oil,
he never cleaned up with the Texas tea,
made his dough in the stickball game,
but George always had strong links to big oil,
family, friends and grand party connections.

So imagine the double take, fresh rush of wonder,
when I heard about his Prairie Chapel Ranch,
otherwise known as the Western White House,
its irregular walls of Texan limestone, numero uno
in eco sustainable features, its geothermal energy
system for heating and cooling, water pumped
through rock from a depth of ninety-one meters
to give twenty degrees comfort all year round.

Boyd Bushman[20]

Nature does not speak English,
Not only that, but if we verbalize it,
we're probably approximating, but not telling the truth.
Math comes close, but it isn't there either.

What Nature tells us is what must be honored.
It has been talking to us on many domains.
I can't talk to all the theoreticians,
because there don't exist theories where I am.

Follow the data and log your data very well
and don't throw it away because you have a theory.
The theory of gravity is just a theory.
Einstein improved on the original theories of Newton
and he was verified by data—and data gave precedence.

No one believed Einstein until the data arrived.
Well, our data's arrived…

20 Found Poem: *from* Nick Cook, *The Hunt for Zero Point;*

Sestina on Pseudo-Science

I breakfast after my sleep is broken.
At the cafe, I order a coffee and torus,
bounce between this world and a dream.
A voice on the radio, whose tone is deep:
The universe is an open loop system.
Prompted to wonder: is this pseudo-science?

I scramble, fall back on my mental science
training, and soon the spell is broken.
Whenever we step out of a closed system,
we enter another (the nature of the poly-torus).
The human race fears an abyss so deep,
as it spirals towards its big bang dream.

The warriors watch their hands as they dream.
Let's find a cure for those blinded by science.
Come along, ensure your pockets are deep.
We assume the bank can't be broken.
The manager's wife has sun in Taurus,
keeps good records of their punting system.

Tourism plans to package the solar system,
a blossoming of the American dream.
I meditate on an image of the hyper-torus.
Quacksalvers often appeal to science.
That radio squawks: the arrow is broken.
Who ever knew our sleep was so deep?

When we join the debts, the queues run deep,
let's weigh the merits of a dipolar system.
It doesn't hurt for symmetry to be broken,
a kaleidoscope that can unscramble a dream.
The skeptical engineer kneels at the feet of science.
But how to find one who can mentor us?

What's become of those young bright auras?
It's been said 'time is energy', now that's deep!
They work at the bench, as we do the science.
To breach the fence around our system,
I'd encourage them to learn how to daydream,
It's all inter-connected, it can't be broken.

Torus thought removes loose ends from a system,
while deep-staters drive me into another lucid dream.
Rest your head, join your scions, back to sleep unbroken.

Unpatented Words[21]

Yesterday, there was a wall of Tesla patents
in the lobby of our Palo Alto headquarters.
That is no longer the case.
They have been removed,

in the spirit of the open source movement,
for the advancement of electric vehicle technology. Tesla Motors
was created to accelerate the advent of sustainable transport.
If we clear a path to the creation of compelling electric vehicles,

but then lay intellectual property landmines behind us to inhibit others,
we are acting in a manner contrary to that goal.
Tesla will not initiate patent lawsuits against anyone
who, in good faith, wants to use our technology.

When I started out with my first company, Zip2,
I thought patents were a good thing
and worked hard to obtain them.
And maybe they were good long ago,

but too often these days they serve merely to stifle progress,
entrench the positions of giant corporations
and enrich those in the legal profession,
rather than the actual inventors.

After Zip2, when I realized that receiving a patent
really just meant that you bought a lottery ticket to a lawsuit,
I avoided them whenever possible.
At Tesla, however, we felt compelled to create patents

out of concern that big car companies would copy our technology
and use their massive manufacturing,
sales and marketing power to overwhelm Tesla.

---

21 Found poem: from statement by Elon Musk, June 12, 2014;

We couldn't have been more wrong.
The unfortunate reality is the opposite: electric car programs
at the major manufacturers are small to non-existent,
constituting an average of far less than 1% of their total vehicle sales.
At best, the large automakers are producing electric cars

with limited range in limited volume.
Some produce no zero emission cars at all.
Given that annual new vehicle production
is approaching 100 million per year

and the global fleet is approximately 2 billion cars,
it is impossible for Tesla to build electric cars fast enough
to address the carbon crisis. By the same token,
it means the market is enormous.

Our true competition is not the small trickle
of non-Tesla electric cars being produced,
but rather the enormous flood of gasoline cars
pouring out of the world's factories every day.

We believe that Tesla, other companies making electric cars,
and the world would all benefit
from a common, rapidly-evolving technology platform.
Technology leadership is not defined by patents,

which history has repeatedly shown
to be small protection indeed against a determined competitor,
but rather by the ability of a company to attract
and motivate the world's most talented engineers.

We believe that applying the open source philosophy
to our patents will strengthen rather than diminish
Tesla's position in this regard.

Up from New York

Nikola Tesla had hobbies and took holidays too,
like when he invited his nephew, so the story goes,
in the summer of '31 to a farmhouse in Buffalo.
In the garage, Tesla unveiled his new Pierce Arrow.

Where the engine should be, sat a cylindrical unit
while the dashboard inside was decorated oddly,
they cruised around town for about a week,
reaching speeds of up to ninety miles per hour
without any need to fill up with petrol.

As the Pierce Arrow idled at a stop sign,
one of a group of young pedestrians pointed:
'Hey, see this car, it has no exhaust fumes!'
His elbow on the door, Tesla's proud nephew
leans out, grins: 'Yeah! We left the motor at home!'

If you want to find the secrets of the universe,
think in terms of energy, frequency and vibration.

Acknowledgments

I wish to thank my lecturers and tutors in the Faculty of Creative Arts at the University of Wollongong for their inspiration and support in the field of writing. In particular, I acknowledge Alan Wearne for the time he devoted to our study of the great range of sonnets written over centuries, and also for covering the phenomenon of found poetry.

I want to thank every financial contributor to this project as it was run on the pozible website. I appreciate your generosity. I marvel at how you agreed to support the project while in possession of such little actual information about the book. I sincerely hope you find satisfaction and enjoyment in your reading of the work.

Thanks also to Julie Bates, Ian Campbell, Gregory Ceff, Jim Hook and Susu Zheng for your valuable feedback in both clarifying the concept of the project and also for feedback on individual poems contained within the work, as well as for the related conversations about language, literature and life. I also thank Rob Millington for our artwork discussions.

Thanks also to my wife, Rubing, for the enormous help she provided in so many ways, including her support during the lockdown period when most of the poems were being prepared, her help with the design of the book cover, and also for the campaign to attract readers and supporters. She was really tireless in her support and I appreciate her efforts more than words can express.

I have drawn on many sources for the details contained within these poems. As online resources operate these days, any of the many details can be tracked by those interested. But a key text for me was Dr Hans Nieper's *Revolution in Technology, Medicine and Society*. In that book, he sprinkled so many breadcrumbs on the trail of this revolutionary new science. I hope this present work serves to continue that trail.

Made in the USA
Monee, IL
04 November 2026

67387162R00066